# AWESOME DINOSAURS

## AN ILLUSTRATED AND INFORMATIVE GUIDE

# AWESOME DINOSAURS

## AN ILLUSTRATED AND INFORMATIVE GUIDE

**George Owen**

# TABLE OF CONTENTS

# INTRODUCTION

Though dinosaurs are extinct today, they have left behind quite a mark on this planet. To this day, scientists want to unveil more information about these prehistoric creatures. The term 'Dinosaur' is a Greek word which signifies 'terrible lizard.' Known for their scary appearances and gigantic shapes, some of these creatures were meat eaters known as carnivores and some were plant eaters called herbivores.

There are many species to take a look at. For instance, many plant-eating species were able to fight against meat eaters. The stegosaurus is one such creature that had spikes on its tail. Meanwhile, the large triceratops was an herbivore that had three horns emerging out of its head.

Dinosaurs roamed the Earth during the Mesozoic Era beginning about 250 million years ago (mya) and ending about 65 mya when they became extinct.

We use a dinosaur as a general term in describing animals that had certain reptile-like characteristics and gained huge proportions. But scientists have narrowed down the definition a bit. A dinosaur is said to have possessed upright limb posture that allow them to use their "hands" for purposes other than running. And a ball-like top to the femur with an open hip socket in the pelvis for a greater range of motion in the back legs. Other characteristics of their bones include special requirements for hands and vertebrae. Some of these defining characteristics could also define humans (even though we are not related except in the remotest sense). Perhaps the most famous of the dinosaurs is Tyrannosaurus Rex whose upright stance is quite pronounced. Rather than four legs, he has two legs and two arms.

Dinosaurs are said to be descended from the Archosaurs (from whom are also descended our Crocodiles and Alligators). Dinosaurs began their existence in the Triassic Period (The Mesozoic Era is divided up into three periods, similar to the way a week is divided into seven days). Many dinosaurs are quite small often weighing less than humans and frequently as small as a human's hand (like the Lagosuchus). The dominant dinosaur type was the prosauropod which was small and lightly built, with a long neck and tail and it fed on both plants and animals.

The Jurassic Period came next and is when large dinosaurs began their rule on Earth. Climatologists believe that the world was still hot and arid at this time, but becoming increasingly wet, so there are more and more jungles and green habitat. The giant sauropods (like the Apatosaurus) began to take over from the prosauropods. Armored dinosaurs like the Stegosaurus also appeared on the scene. The armor is thought to have been a way for plant eaters to protect themselves from the meat-eating dinosaurs like the Allosaurus. Brachiosaurus, which is the tallest of dinosaur also munched its way through this period.

Very high sea levels and humid, warm temperatures marked the subsequent Cretacious Period. This has been marked as the heyday of the dinosaurs. Many new species like the Spinosaurus and the Parasaura-lophus, appeared, perhaps in response to the evolution of flowering plants. In the oceans the Plesiosauruses made their presence felt. One of the most remarkable dinosaurs of the period was Triceratops, where a large neck frill provided both protection and an opportunity for display. Some plant-eaters also developed armor on their backs and even club tails that allowed them to counter attack against the pursuing carnivores.

The Cretacious Period ended with a mass extinction that completely wiped out the dinosaurs. There are numerous explanations accounting for their extinction. No one is quite certain as to the cause. Some conjecture that an asteroid may have struck the

Earth or that some other event came along to create a severe climate change. Some believe that such a thorough destruction of a whole group of creatures may have been the product of a virulent virus that evolved and was passed along by flying creatures as well as insects, which were also becoming more and more pervasive during this time in geologic history. Another explanation offered is the massive volcanic eruption. This implies that volcanic eruptions were happening even in the age of dinosaurs. Upon eruption, the Earth is covered with sulfur dioxide that blankets the Earth's surface. This cloud blocks the heat and light of the Sun that was supposed to enter the Earth. This made the Earth's temperature to plummet and it killed all the living things during those times.

There are some misconceptions about pteranodons being flying dinosaurs. Pteranodons were not dinosaurs per se; they were separate ranch of the family tree. Flying dinosaurs though did exist and may have descended from small, fast running maniraptor dinosaurs. They would eventually become birds. The first bird/dinosaur is thought to be the Archaeopteryx. Because of their size Mammoths are sometimes lumped together with dinosaurs. However, Mammoths lived much later and were mammals.

Over time, scientists have identified over thousand species of dinosaurs across the globe. In fact, they have discovered the fossils of these creatures on every continent. However, they did not restrain their studies to this number and are still exploring new fossils to prolong their list of this known species.

Scientists have continuously learnt more about dinosaur every day but evidence aboput these creature is millions years old. At this point, it is important to know what most of the creatures looked like other than in outline. Feathers, skin, and flesh seldom survived the ravages of time to our own geologic era. Also, the number of skeletons available for study, though growing, is not vast. Thus, much of what is thought about dinosaurs is not actually "known" about dinosaurs. Because of this, there is still much debate over such issues as whether dinosaurs were warm or cold-blooded. How intelligent were they? Did they have a herd instinct? Were they generally drab like lizards or bright and colorful like birds? The answers to these questions are much likely to find answer as science continues to quest into the past, into the world of the dinosaurs.

Although I'm neither a palaentologist nor a scientist, I'm a huge and lifelong fan of these creatures and I'm always hungry for new discoveries about these fantastic beasts. This guide is intended to provide easy access to information about the best-known dinosaurs; they are likely to be encountered in museums.

# TYRANNOSAURUS REX

The Tyrannosaurus is often abbreviated as T-Rex, it is generally considered as the king of dinosaurs. As a matter of fact, the Rex in its name is latin for "King" while "Tyrannosaurus" means "Tyrant Lizard." When put all that together, dinosaur name is translated to "Tyrant Lizard King." Their fossils were found in the present western part of North America. It lived during the Cretaceous Period alongside some other famous dinosaurs like the Triceratops and Hadrosaurus both of which probably acted as food sources for this large and hungry Tyrant Lizard. Below are some of the striking features of this giant creature that made it different from the other species of dinosaurs.

They are believed to be one of the largest creatures that existed on Earth. Bipedal carnivore, Tyrannosaurs were huge in size and it measured approximately 40 feet long and about 13 feet from the hip. The female T-Rex had enormous pelvic structure probably for the passage of eggs. Detailed research though has doubt about its exact mass but it's believed to vary between 5.2 to about 6.6 US tons.

They had an S-shaped neck structure, though short, but capable enough to support the enormous skull and the heavy long tail. Their skulls were about 5 feet long and extremely wide with a narrow nozzle that gave them the unusual ability to have a binocular vision. Due to the powerful nature of the skull bones, they were extremely able in hunting their preys without failure. Research has unveiled the fact that they had the strongest biting ability that any creature could have and a comparable study has shown that it could be imagined to be about 15 times of that of an African lion.

They undoubtedly had powerful forelimbs with two clawed fingers with a metacarpal characterizing the residue of the third finger. Their hind limbs were comparatively huge and were solely responsible for balancing its massive structure. Though enormous in size, T-Rex had no movement difficulty since most of the bones in the skeleton were hollow, reducing a considerable amount of its mass without affecting its strength.

The unbelievable growth curve these creatures had. A juvenile weighted 2 US tons till 14 years age. After that phase, their body mass changed hugely and it is believed to be about an increase of 0.66 US tons per year. Now you know what makes T-Rex one of the largest creatures and definitely king of dinosaurs. Though their lifespan was not that huge, it was only limited to about six years after reaching its sexual development.

Traditionally, Tyrannosaurus Rex has been depicted as a fierce and active predator, however, in recent years, some scientists have challenged this view. Those scientists who support the traditional view of Tyrannosaurus point to the relatively large areas of the animal's brain devoted to sight and vision (which would have been useful for hunting), and the arrangement of the animal's eyes (which are for-ward-facing and suggestive of binocular vision, also useful for hunting). On the other hand, those scientists who have challenged the traditional view, argue that Tyrannosaurus was probably relatively slow moving and possibly lived as a scavenger feeding on carcasses killed by other animals.

# TRICERATOPS

If you were to walk up to a stranger and ask them to name three dinosaurs, they would almost be guaranteed to name the Triceratops. The fact is, most people have a fond place for the Triceratops, a unique dinosaur that is as iconic as the Tyrannosaurus Rex and a personal favorite of mine.

Triceratops was around 25 feet long, similar in length to a Stegosaurus, however, it lived in the late Cretaceous period, around 68 million to 65 million years ago. Unfortunately for Triceratops, we all know who else lived at this time, the Tyrannosaurus Rex. To make matters worse, the films could have been right in their frequent depiction of a dueling fight between these two giants, as fossils show they lived in the same regions of the planet, such as Wyoming, Colorado, and Montana to name a few.

Unlike Tyrannosaurus, Triceratops was a herbivore and ate fibrous plants, which at the time would have been in abundance. After eating a good meal, a fully grown adult would have weighed in at around 5 tons, that's just over 3 cars weight to you and me. When considering their shape and weight, it's doubtful that they would have been that quick on their feet.

The most distinctive feature to a Triceratops was the skull shape. They had a pair of long brow horns and a shorter nasal horn. In addition to the horns was a large frill on the back of their head. The two large horns could have grown up to three feet long and the frill could measure up to seven feet wide.

The features on a Triceratops could have provided fantastic protective weapons against any predator that decided they fancied a fight. It's possible that our plant eater would have used its great weight and charged at on comers, much in the way a Rhino does today, however, some experts now believe that these characteristics were only really used in courtship and not as defensive weapons.

Like many species of dinosaur, there has been much debate in relation to how many different types existed, such as today's Black and White Rhinos. At one stage, experts thought that there could be as many as sixteen different variations belonging to the

Triceratops family, however, more modern day thinking suggests these were just variations within individuals and it was more likely that there were just one or two valid species.

A Triceratops would have been an awesome sight and many films, books and toy ranges, such as Papo, Safari or Bullyland, have provided wonderful depictions. Always a popular favorite amongst children and adults alike the Triceratops is one of the unique creatures to have ever walked the Earth.

# VELOCIRAPTOR

The Velociraptor was made famous by the film Jurassic Park and the subsequent sequels. Now the term Raptor is used in general dinosaur conversation as often as T-Rex and Triceratops, however, the image that these films created, of a tall, smart, quick and dangerous dino was in fact somewhat different to the reality.

Velociraptor, or Velo as we're now going to refer to them, lived in the late Cretaceous period, although some speculate it would have been closer to the late Santonian period or even earlier, some 85 million to 80 million years ago. As a meat-eater, found in parts of Russia, Mongolia, and China, they were roughly 6 feet in length (not height) and were no larger than a wolf. While there's no doubt they would have been swift on foot and probably smart, they would not have had the speed of a Cheetah or the intelligence the films suggest.

Velos had sharp teeth and large eyes and of course, the famously distinguishable sickle-like claws on the second toe of its hind feet. It was in the early 1960's that experts realized the importance of these claws and pieced together that these assets would have been used as slashing weapons. After capturing a light weight prey, and in some cases relatively large prey, the Velos would have held onto their victim with their long fingers in order to hold them down and use the sickle-like claws to kill their victim.

In 1971, famous specimen was found in the Gobi Desert, where the Velo was found still holding onto a Protoceratops. These creatures died at the same time, locked in a battle to the end.

Recent finding has indicated that Velos may have been feathered in some manner, which is perhaps an early example of the beginnings of the evolutionary change to birds.

Despite the differences between the Jurassic Park version of a Velociraptor and the ones that walked the Earth some 80 million years ago, there is no doubt that these creatures would have been a voracious predator and deserve recognition.

Today, Velos appear on many toy merchandise and many top brands have created their own image of this dinosaur, such as Papo, Bullyland, and Safari. While these brands have all created their own look for a Velo, they are all distinguishably different, for example, some have feathers and others are sized smaller.

Though there are several possibilities to explain the massive dying out of the dinosaurs 65 million years ago, the most popular today is the meteorite or asteroid impact theory. It is said that this cosmic visitation could have slammed a crater on the planet 200 kilometers across, vaporizing rock and flinging dense clouds of dust and water droplets into the atmosphere. Strong winds dragging these particles could have shrouded the planet causing dark global storms, torrential rains that caused floods to drown and kill the dinosaurs, including the Velociraptors. Virtually all Velociraptor fossils indicate a swift and merciless death by water.

# STEGOSAURUS

Stegosaurus is a popular dinosaur amongst dinosaur enthusiasts, both young and old. There are many reasons for their popularity which include their unusual distinctive shape and the fact they are considered to be nice dinosaurs.

To put a bit of meat on the bones, the Stegosaurus lived during the late Jurassic period, around 156 to 145 million years ago. They had a smallish head in relation to their overall size which leads to a large bulky body. The body, in turn, was carried on short legs. They had heavily plated backs and have been found all over the world. In fact, Stegosaurus was the first plate-backed dino ever found.

Stegosaurus is quite simple to recognize. It had a small head, a bulky body, and a tail. Stegosaurus is usually depicted as a quadruped (walking on four legs), but it probably could have stood in a bipedal stance on its powerful hind legs when browsing from the branches of trees. The most recognizable feature of Stegosaurus were the bony plates on its back - these are usually depicted as standing upright and were arranged in pairs or a zigzag fashion - however, there are some scientists who believe the plates may have lain flat on the animal's skin. As well as the bony plates on its back, Stegosaurus also had a number of spikes (the number varied depending on the particular species) at the end of its tail.

Believe it or not, despite what all of the films show, the back plates still remain an area of controversy. Firstly, either Stego, as I'm now calling them, had a single row of plates down their spine, or as an alternative theory, they had two rows that were not lined up against one another. Secondly, we once thought that the larger plates were placed on the back for armored protection, however, expert have suggested recently that they were there in order to help control body temperature and fossils collections would support this assumption.

Regardless of the back plate positioning or the actual use of those plates, Stego would have looked a lot larger with them, than without them, and as an adult would have reached around 25 feet long, which may well have put those pesky, cunning meat-eaters at bay.

Meat-eaters would have been a problem for the poor old Stego, specifically when you consider Stego's brain size. Yes, the old theory that a dinosaur had a brain the size of a ping pong ball is probably referring to the Stegosaurus. Weighing in around 70 grams, Stego's brain was, in fact, a wee bit smaller than a ping pong ball, making you have to think that they were at a great disadvantage. Of course, all this said, you have to remember that they were found all over the world in large numbers, so they seemed to survive just fine.

Today, there are plenty of good examples of Stegosaurus skeletons in museums around the world and I would suggest they are worth a look. Further, failing this will always feature in books, be presented on the internet, and without a doubt make up a dinosaur toy collection, so there's always an opportunity to admire this impressive dino.

# SPINOSAURUS

Spinosaurus have no same recognition as a Tyrannosaurus, specifically when discussing general popularity. In fact, some actually haven't heard of the dinosaur Spinosaurus in order to form a comparison. Despite everything we know about the T-Rex, the Spinosaurus may well have been the largest meat-eater to have ever walked the Earth.

To put this into context, the vertebrae of a Tyrannosaurus Rex measured roughly 6.3 inches long, while the Spinosaurus measured closer to 8.3 inches long.

The most obvious feature of Spinosaurus is the large fin on is back, sometimes also referred to as"sail". This fin was probably used for temperature regulation - the animal could face it towards or away from the sun depending on whether it wanted to heat up or cool down. It is from this fin that the animal gets its name - Spinosaurus means "spine lizard".

Another notable feature of Spinosaurus is the relatively large arms - compare for example the puny arms of Tyrannosaurus Rex. Perhaps it's thought that these large arms may have allowed Spinosaurus to some-times walk on all fours, rather than on just its hind legs.

The Spinosaurus was around 50 feet long, although many believe they would have been even larger. They lived in the late Cretaceous period, around 97 to 95 million years ago and inhabited the swamps of Northern Africa, with fossils having been found mainly in parts of Egypt and Morocco.

With a face similar to a crocodile, only vastly larger, Spinosaurus would have primarily lived on a diet of fish and small herbivore dinosaurs. However, they weren't against general scavenging around for anything that was available. Evidence showed they would have lived in water and on land, which would have provided ample room for natural food resources. Besides its massive size and peculiar face, the enormous sail on its back is what made this dinosaur really stand out. The spines alone were over 5 feet long and were connected by skin, much like the earlier Dimetrodon. As with most dinosaurs and the scientific world, controversy arrives with many debate over the purpose of the sail. Most believe its main function was to regulate the body temperature by absorbing heat, while others believe it was used in courtship displays. Either way, most are in agreement that the sail would have been brightly colored, similar to the fins displayed on some modern-day reptiles. Interestingly enough, if the spines on a Spinosaurus were acting as radiators, it may prove that not all dinosaurs were warm-blooded.

Unfortunately, with the first set of fossil remains being lost to us in the bombing raids over Munich during World War 2, there are not any fully complete skeletons of Spinosaurus to view in museums. In addition, Spinosaurus hasn't yet featured in many dinosaur films, although he does get to star in a Jurassic Park sequel. This said, you can still see plenty of quality recreations through the internet, books and toy ranges.

# ARCHAEOPTERYX

**Archaeopteryx - Fossilised Pigment Structures Suggest First Bird was Black**

A team of international scientists studying a single fossilized feather ascribed to the Jurassic "missing link" Archaeopteryx has found prehistoric pigment structures. These structures, known as melanosomes suggest the color of this Jurassic creature and also provide an insight into whether or not this bird from the time of the dinosaurs was a strong flier.

Using a powerful, scanning electron microscope located at the Carl Zeiss laboratory (Germany), the researchers were able to identify hundreds of tube-shaped structures towards the tip of the single, fossil feather they studied. The shape and density of the melanosomes give an indication of the color of the creature when it was flying around a Jurassic lagoon, approximately 150 million years ago.

## Melanosomes Indicate Archaeopteryx had Black Feathers

Archaeopteryx was formally named and described in 1861. This crow-sized creature shows bird-like as well as reptilian characteristics. Fossils of Archaeopteryx (A. lithographica) show feathers, but also reveal teeth in the jaws and a long tail, anatomical features associated with dinosaurs. Just ten specimens of this creature have been found, all from the Solnhofen sediments located in southern Germany. The Solnhofen strata consist of deposits of finely layered, fine-grained limestone. These limestones were laid down at the bottom of a shallow lagoon cut off from the Tethys Ocean which covered much of Europe during the Late Jurassic, by a large reef. The bed of the lagoon could not support life. Animal and plant remain that fell into the lagoon and settled on the bottom were not scavenged by organisms and so they had a chance of being preserved almost intact. Thanks to the exceptional quality of the Archaeopteryx fossils, and this single feather, scientists have determined that the tips of the wings were colored black.

## Scans Reveal Structure of Archaeopteryx Wing

Scientists have discovered that the layout of the feathers on the wings and body of Archaeopteryx were arranged in the same pattern as feathers on the wings of modern birds. In addition to the discovery, these feathers were asymmetrical in shape, the same shape as the feathers on modern birds. Asymmetrical feathers are essential to permit powered flight, in the same way, that an aircraft's wing is shaped to create lift.

This new study, adds weight to the "powered flight theory". An examination of the tiny, hook-like structures known as barbules on the wing feather shows that the microscopic structure of the Archaeopteryx feather was identical to that seen in examinations of the feathers of modern birds.

## Archaeopteryx may have been a Strong Flier

This new evidence suggests that this primitive bird, a creature that lived during the time of the dinosaurs, may have been quite a strong flier. Being a powerful flier may have been a very useful survival strategy with lots of agile dinosaurs around keen to make Archaeopteryx their dinner.

# BRACHIOSAURUS

The Brachiosaurus was a gigantic dinosaur that has been popularized by many films such as Jurassic Park. Film depictions of this dino have steadily improved over time, although there has always been the common theme of a slow, gentle and friendly creature which most kids over 5 years of age could probably describe in reasonable detail.

At an impressive 82 feet long, 52 feet high and weighing approximately 38,5 US tons, Brac, as we're now going to refer to him, was one of the true heavyweights this planet has ever seen. There have been claims of larger dinosaurs being located in recent years, but for a very long time experts believed this to be the largest of all land dinosaurs.

Bracs lived in the late Jurassic period, around 156 to 145 million years ago. Specimens have been discovered in Africa and North America, where they would have fed on lush vegetation. There are some differences between the two species, specifically the potential height difference, however they are basically the same dinosaur.

In some respect, Bracs were a bit like a modern-day giraffe, size difference aside. Bracs had a small skull which was perched on a very long, potentially large muscled neck that sat on a large, wide body with long front legs and slightly shorter rear legs, which again were potentially built up. To compensate for the neck was a long tail. Considering the frame and size of Bracs, they were actually lighter than experts had originally thought.

There were many predators looming in the late Jurassic period which has left experts debating whether Bracs could have found themselves on the menu. For a very long time, it has always been argued that Bracs were just too large for any predator to attack, for as large as the predators were they simply weren't anywhere near the size of these large herbivores. There is a growing community that now believes that some of these large predators in fact hunted in packs, not dissimilar to the lions of today. Unfortunately, despite some interesting finds that would support this new thinking, there just isn't enough evidence to say that Bracs were in any kind of danger.

Brachiosaurus was a true wonder of the world. One of the largest creatures to have ever walked this planet. The internet and books are a great source of information and you can see various pictures of how experts believed Bracs would have looked.

There has been a tremendous amount of scientific research carried out on Brachiosaur fossil material over the last few years, indeed there is as strong argument to split the African and American Brachio-sauridae fossils into separate genera. Recent studies have suggested that these dinosaurs were actually much lighter than previously thought, but Papo has opted to give their replica a very thick neck and a large body supported by very strong pillar-like back legs. This reflects a more traditional view of the Brachiosauridae, the model gives the impression of a very powerful animal, one of the largest land-living animals known to science.

# ALLOSAURUS

The Allosaurus possibly doesn't get the same recognition as other large meat-eaters such as Tyrannosaurus Rex, but make no mistake this was an awesome predator that deserves people's attention. This dinosaur was probably the most common carnivore of the late Jurassic period due to a large number of fossil remains that have been collected. Unfortunately, there have not been enough complete specimens found, as there are possibly two large carnivores of the period that go under the same name, but not enough evidence is available to distinguish if they are specimens of the same species.

Allo, as we're going to refer to the Allosaurus from here on, was around 30 feet to 40 feet long, and lived in the late Jurassic period around 156 million to 145 million years ago. Fossil remains have been located in areas of Colorado, Utah, and Wyoming to name a few.

Allo, had a large three feet long head and its lower jaw would have been hinged on, allowing it to wolf down large chunks of meat in one go. Powerful large hind legs that would have been heavily muscled gave it the ability to run very fast, around 20 miles plus per hour in relatively short bursts.

The arms of Allo or forelimbs were much shorter than the hind legs, however, they had three very sharp claws on them which would have been used to hold its prey down. To add menace, Allo had short horns located above its eyes.

There has been speculation as to Allo's prey, as while there were many dinosaurs that Allo could easily kill, one has to remember that they lived at a time when the Sauropods were in abundance, Sauropods such as a Diplodocus that would have been massive in comparison. Some experts believe that Allo may well have attacked these large dinos, certainly the young, however, this would not have been single-handedly, but instead in a pack, similar to Lions today. Later speculation has suggested that Tyrannosaurus Rex would have acted in a similar fashion.

Allosaurus has not yet featured in many Dinosaur films, in fact, none that come to mind, but the fact remains these dinos were large predators that would have created a storm if T-Rex and others hadn't been around.

Today, there is some debate about Allosaurus' lifestyle. Some scientists believe that Allosaurus may have been too large and heavy to chase down prey, and therefore probably got most of its meat by scavenging (for example, scavenging the carcasses of animals killed by other carnivorous dinosaurs). Other scientists, however, have formed that view that, despite its large size, Allosaurus might actually have been surprisingly agile and could perhaps even have hunted in packs. Scientists of the later view, therefore, believe that it was quite possible for Allosaurus to be able to bring down large herbivorous sauropod dinosaurs such as Apatosaurus (also known as Brontosaurus) and Diplodocus.

# ANKYLOSAURUS

A well-known armored dinosaur, Ankylosaurus was once the victim of misidentification. While the image today of this remarkable dino is unmistakable, Ankly, as we're now going to refer to them, was presented in a very different form. American Museum of Natural History prospector, Barnum Brown, thought that Ankly was some sort of Stegosaurus and pieced a skeleton together with a high arching back and circular armored patterns. As we now know today, he was very wrong as Ankly looked nothing like that reconstruction.

Ankly lived in the late Cretaceous period, which was around 70 million to 65 million years ago. Fossil remains have been found in Montana, Wyoming, and Alberta. Unfortunately, Ankly bones are pretty rare and only three good quality specimens have been found. Experts have speculated that this is because Ankly probably lived on higher ground and not near the river deltas where most fossils have been located. The three full skeletons that were found were probably either washed downstream or may have been older animals who lost their way.

A plant eater who would have eaten vast amounts, Ankly would have reached lengths of over 25 feet as a fully grown adult. With a stocky build, but without much height, one could be mistaken for thinking this dino was bred for fighting as Ankly was a heavily armored creature that could do serious damage if it needed. The armor was made from bone pieces on the outside of the skin. In fact, their armor was so heavy that even their eyelids had protective plates.

The most impressive feature of an Ankylosaurus was its tail, as Ankly had armored plates that ran down its vertebrae and onto the tail itself. If the plates didn't act as enough protection then the vertebrae of the back one-third of the tail could overlap, turn it into a solid handle which had several plates at the end acting as a club. These plates fused together into a single unit with a further two plates on either side.

That's right, as much as Ankly was protected from predators, it also had a well-equipped weapon that could do major damage.

Ankylosaurus is yet another fantastic example of a unique dinosaur that wandered this planet long before mankind. They may not stand out in people's imagination as much as a Tyrannosaurus Rex or the lovable Triceratops, however, there is no denying Ankylosaurus was a truly unmistakable dinosaur.

Ankylosaurus was named by Barnum Brown, one of the greatest dinosaur hunters of the 20th century. The first fossils of the animal were found by a team led by Brown in 1906. Barnum Brown chose the name "Ankylosaurus" in 1908, and like most scientific names of animals, it is derived from ancient Greek - "Ankylosaurus" means "stiffened lizard", and refers to the fact that many bones in the animal are fused together.

# DILOPHOSAURUS

Not necessarily a familiar dinosaur name for many people, however, the Dilophosaurus had a good role in the first Jurassic Park film. That's right, Dilophosaurus was the small dinosaur that happily spat poison at the ultimate bad guy of the Park as he tried to steal Dinosaur DNA for personal gain. While the film did a good job in providing a great image and personality to this dino, I'm afraid there isn't anything from the fossil collections to suggest that Dilophosaurus did, in fact, spit poison.

Dilo, as we're now going to refer to them, lived in the early Jurassic period, around 208 million to 194 million years ago. Fossil remains have been found in Arizona mainly and confirm that Dilo would have been a meat-eating dinosaur.

As the film suggested, Dilo had an unusual pair of highly arched crests on the top of its head. These crests may well have been colorful or even patterned. Experts believe Dilo's may well have used these crests for display purposes in courtship, although a few individuals have suggested they may have been used as a signal deterrent against large carnivores of the time.

Fossil collections and multiple small fossilized footprint tracks, have lead individuals to believe that Dilo's may well have complex social behaviors which might have included herding and seeking out food together in an organized way.

A standout feature of the Dilo's in the film Jurassic Park was the elaborate fan around their neck. The film indicated that they would raise the fan neck dressing when they were attacking or spitting poison at prey. Similarly to the poison position, which I stated earlier, there is no evidence to suggest that they had a fanned neck.

Dilo's had long slender teeth which were perched in their large head. Dilo's skulls are considered large in relation to their body size. Experts have suggested that they would have lacked a strong bite due the fragile fossil remains and the skull proportions. The lack of bite power would indicate that Dilo's were in fact scavengers which plucked at meat and used grasping hands and feet to tear the flesh rather than their jaws.

Dilophosaurus was an amazing dinosaur. Where this dino lacked size and power it made up for in unique features.

# PTERANODON

The Pteranodon was actually not a dinosaur. It was a flying reptile-like creature that developed at the same time as the dinosaur, possibly from the same ancestor. It was classed as the Pterosaurs.

Like the other Pterosaurs, the Pteranodon walked upright and had straight ankles. It had hollow bones that makes it lighter for flying. The wings were a development that came as a modification of the Pterosaur's hands. The fourth finger became extremely elongated and acted as a frame for the wing membrane which allowed for flight or at least gliding.

The other three fingers remained present and stuck out about half way down the wing.

They had no teeth and were tail-less generally (if the very short tail is disregarded). The Pteranodons lived during the late Cretaceous Period. They had a significant growth rate, with wing spreads of 25 feet or more! It is believed that Pteranodons grew even larger towards the late Cretaceous Period.

The flying Pteranodons did not have tails or teeth like their predecessors, the pterosaurs, who roamed the Earth (mainly Europe) during the Jurassic era and were usually of much larger size. The wing membranes of the Pteranodons were no longer in connection with the lower legs.

These flying reptiles were presumably warm blooded and had a narrow shield of hair all over their bodies. They spent most of their time flying over the ocean's waters hunting for their preys. They used to feed on fish and other marine animals and played a role in the Cretaceous Inland Sea, pretty much like other seabirds such as nowadays' albatross and pelican.

With its long beak, the Pteranodon may have skidded along the surface of lakes or bays to scoop up water creatures. The large crest sticking out of the back of its head acted as a counterweight to this heavy beak allowing it to remain balanced in flight. Pteranodons are thought to have had webbed feet that aid in swimming. They probably lived in coastal regions living on fish. Some paleontologists believe that they were warm-blooded and coated with fur or possibly feathers.

The Pteranodon could reach six feet in length and weighed about 35 pounds. Its wingspan could reach as wide as 23 feet. Remains of this creature were found in the southwest of the United States and were first described by Othniel C. Marsh in 1876, in the form of fragments of wing bones much larger than the pterodactyl remains previously discovered in Europe. It lived about 85 to 75 million years ago during the Cretaceous Period of the Mesozoic Era. The Pteranodons were highly developed special species of pterosaurs.

# MOSASAUR

Mosasaurs are a group of extinct marine reptiles. They were powerful swimmers with long stream-lined snake-like bodies (although they did have four limbs all finned, and possibly a finned tail), and ate fish, turtles, sea urchins, and shellfish including molluscs. The smallest known mosasaur was about 10 feet (3 meters) long, but the largest grew as long as 57 feet (17.5 meters).

Mosasaurs were not dinosaurs, but were lepidosaurs (reptiles with overlapping scales, the group that includes lizards, snakes, and sphenodontia such as the tuatara).

Mosasaur is believed to have evolved from aigialo-saurus, which was semi-aquatic lizard that lived during the early Cretaceous period, and who is believed to be related to monitor lizards.

Mosasaurs are considered one of the Great Marine Reptiles that ruled the seas during the Cretaceous period. Other great marine reptiles at that time include the dolphin-like ichthyosaurs, the long-necked plesiosaurs, and the short-necked pliosaurs. Luckily for us, all Great Marine Reptiles became extinct at the end of the Cretaceous period 65 million years ago.

Once mosasaurs returned to the seas in the Cretaceous, around 100 million years ago, they rapidly diversified. Numerous subfamilies, genera, and species appeared near-globally throughout the Cretaceous. They even expanded into freshwater environments. In 2012, Laszlo Makadi published a paper on the discovery of a freshwater mosasaur from Hungary that lived in the rivers, similar to the freshwater river dolphins today.

In 1869, Edward D. Cope suggested that Mosasaurs and snakes share a common marine ancestor. This idea was based on the similarities observed in Mosasaur and snake jaws, the reduced limbs, and the fact that Mosasaurs may have moved in a similar way to snakes. In the 1990s, the discovery of fossils of early snakes with vestigial limbs in marine sediments seemed to provide support for this hypothesis. However, more recently, other early snake fossils have been found, and since these show animals with hind limbs and an apparently burrowing lifestyle, some doubt has been cast on the idea that Mosasaurs and snakes shared a common ancestor.

Mosasaurs first evolved was during early or middle Cretaceous period, perhaps around 96 million years ago. In the last 20 million years of the Cretaceous, following the extinction of Ichthyosaurs, they became the dominant predators. However, all Mosasaurs died out during the mass extinction at the end of the Cretaceous period.

The first publicized discovery of a Mosasaur fossil occurred in 1778. A fossil was found in a limestone quarry in 1780, near the city of Maastricht in Holland. It was not scientifically named until later, the name Mosasaur means "Meuse lizard", and refers to the nearby Meuse River was eventually given. Subsequently, other fossils which had been found earlier in the same area, and had been on display since around 1770, were also identified as being from a Mosasaur. Since then, other Mosasaur fossils have been found in many other countries around the world, including Australia, Canada, Denmark, Mexico, New Zealand, Peru, Sweden, and the United States, as well as in Africa and off the coast of Antarctica.

# PACHYCEPHALOSAURUS

The Pachycephalosaurus was a bone-headed dinosaur which was a rare breed indeed. Some may know or recognize this dinosaur from the Lost World movie, which was the sequel to Jurassic Park. The Pachycephalosaurus was the shortish dinosaur that the hunting party snared, however, it was at the cost of a big struggle that saw him do a lot of damage to their vehicles. Later in the film, and thanks to one of the lead characters, we saw that he was able to get away unscathed.

The Pa, as we're going to refer to them, lived in the late Cretaceous period, around 68 million to 65 million years ago. Fossil remains have been found in Montana, Wyoming, South Dakota and even as far as

Alaska. Pa was a plant-eater and would have dug its vegetation up from under the ground.

The large skull which held the bone shape on its roof would have measured around 25 inches in length for an adult. There are various theories for the bone head, some experts have surmised that they would have used this feature to head-butt rival males in contests, which could have either been territorial or in the mating season. Other experts have rubbished this idea as they believe the bone wasn't strong enough for this type of use and they would simply end up killing themselves. The plausible explanation today is that they would have head-butted other dinosaur flanks in order to ward them away.

Pa had bony spikes on the end of its snout which would have added a layer of menace in appearance, however, it is commonly believed that this would have been used as an advanced tool for digging up food and vegetation.

Pa's would have grown to around 14 to 16 feet long (not high) and would have only weighed between 900 pounds to 1000 pounds, which in terms of dinosaurs wasn't that large. It is thought, due to skull shape and the type of dinosaur Pa was, that they would have had a very good sense of smell which again was probably used in their search for food.

Pa's was probably a herding creature, running in large packs. Incidentally, running would have been their only real form of defense as there were some large predators that wandered the territory at that time.

Pachycephalosaurus was another example of a remarkable dinosaur surviving in a very dangerous time. There are some great images of Pachycephalosaurus recreations which can be found in dinosaur books and on-line.

# COMPSOGNATHUS

Compsognathus live during the late Jurassic Period 156-145 million years ago in what is now Europe. It was lightly built and a fast runner.

Not very many specimen have been found from this dinosaur. Compsognathus shares the same locality and the same limestone deposits as the famous Archaeopteryx, and it has been sometimes confused with it. The two animals have similar sizes and skeletons. But there are some differences, one of the most obvious is that Archaeopteryx had feathers, and Compsognathus did not.

Compsognathus was only the size of a chicken. It is believed to have been one of the smallest known dinosaurs. Compsognathus had two long and thin legs and feet with three toes each. The Compsognathus' head was small and pointed and it had small but sharp teeth. The other distinguishing features of the Compsognathus include the hollow bone structure and a long and very flexible neck.

Compsognathus had short arms, both with two clawed fingers. Its long tail offered the dinosaur balance and stability during fast turns. Compsognathus or 'pretty jaw' was named in 1859, by Johann A. Wagner. The dinosaur was a carnivore that caught and ate smaller animals, insects, and lizards. They fed on the lizard called Bavarisaurus that was known to be a very fast runner.

Compsognathus was between 28 inches and 4.6 feet long and stood 10 inches or 26 cm tall, at the hips. It weighed 6.5 pounds or 3 kg. Fossils of Compsognathus was discovered in Germany and France, they belonged to the superfamily of maniraptoriform, which indicates that they were advanced coeluro-saurs with fused wrist bones.

When Compsognathus was discovered, paleontologist Andreas Wagner did not think it was a dinosaur because of its size. People at that time were used to the idea that dinosaurs were large animals. Now we know that dinosaurs came in all sizes.

# APATOSAURUS (BRONTOSAURUS)

This dinosaur is widely known as Brontosaurus (meaning "thunder lizard"), however, the correct scientific name is Apatosaurus (meaning "deceptive lizard"). This is because when Othniel C. Marsh discovered the first two fossils in 1877 and 1879, he did not realize that they were adult and juvenile examples of the same species, and hence gave the two fossils different names. Subsequently, it has been realized, that both of these fossils were after all from the same species, so only one name should be used: Apatosaurus is the correct name because it was the first name used.

Apatosaurus was a massive herbivore (plant-eater) that lived during the late Jurassic period 157 to 146 million years ago in western North America. It is a whip-tailed, long-necked dinosaur, measuring roughly 70 up to 90 feet long and 15 feet tall around the hips. The long neck consists of 15 vertebrae and hollow backbones. Its long "whip-like" tail is approximately 50 feet long. It also has "peg-like" teeth in the frontal part of its jaws. It has 4 "column-like" massive legs, with the hind legs being larger than the ones in front. The nostrils are on the upper part of its head.

Its 20-foot-long neck supported a rather small head and its brain was about the size of a large apple. Two low ridges ran the length of its backbone, from the base of its skull almost to the tip of its tail. These spines supported the ligaments that held up the neck and tail. Apatosaurus' 30-foot whip-like tail and its enormous size were probably its means of self-defense. This dinosaur lived in western North America, where most specimens have been recovered.

When it was first discovered, many scientists be-lieved that Apatosaurus was so huge that it could not have supported its weight on dry land, and therefore must have lived partially submerged in water. If you look at older books about dinosaurs, you may often see Apatosaurus depicted in this way. More recent research and discoveries (including the discovery of fossilized footprints), however, suggests that Apatosaurus probably did live on dry land, and probably was a grazing animal that lived in herds.

Other interesting facts about Apatosaurus include:

1. Apatosaurus was a close relative of Diplodocus. Apatosaurus was not as long as its relative, but it was bulkier.
2. Apatosaurus could have reared up on its front legs and brought them down with crushing force on an attacker or used its tail as a whip in defense.

Apatosaurus and some of the other large sauropods would have required having large, powerful hearts, to sustain very high blood pressure, in order to deliver blood to their small brains. The head of Apatosaurus was held high above its heart. This presents a problem in blood flow physiology. To sustain this tremendous high blood pressure, an annimal of this size would have required to have wide, muscular blood vessels with many valves to prevent backflow of blood. Apatosaurus' blood pressure was probably over 400 mm Mercury, three or four times as high as ours.

# PARASAUROLOPHUS

A reasonably well-known and easily identifiable dinosaur, the Parasaurolophus made a brief appearance in the original Jurassic Park film, seen herding in front of the lake near the beginning of the film. They again featured in another brief outing in the follow-up to Jurassic Park, 'The Lost World', where one unlucky Para, as we are now going to refer to them was unsuccessful in its attempts to evade capture. Later it was seen escaping thanks to a little help from Vince Vaughn.

The Para's lived in the Late Cretaceous period, around 76 million to 65 million years ago. They would have survived on plants, as they were strict herbivores. Fossil remains have been located in Alberta, Utah and New Mexico and they were the most highly derived member of the Lambeosaurine family, which were a large group of duck-billed dinosaurs. Unfortunately, while we have some good examples of this dinosaur in the fossil collections there haven't been many found, in fact, out of their particular family group they are at the bottom of the list for recovered fossils.

From the few fossils collected, we know that Para's were around 30 feet in length and experts believe that they were at home walking on either all fours or on two legs. A fully grown adult would have weighed in at over 2 tons, roughly a car and a half.

The most striking feature was the Para's head, which was crested and curved over at the end. This crest was longer than the entire skull and housed hollow tubes that ran to the end and looped back again. Typically, as with most dinosaurs, there has been much debate, this time it's the purpose of the Para's crest that has drawn the conversations and left us with two main schools of thought. Originally, experts thought this shape was formed to assist with swimming underwater, a snorkel type device, however, this has been much discredited as it would have required nostrils at the end of the crest which clearly wasn't the case. Current thinking suggests that the crest simply assisted in creating a low-frequency sound which would have acted as a form of dino communication. The sounds would have carried a long way, similar to that of elephants today.

Para's probably had a very good sense of smell and it would have been this that really assisted them in survival situations as they had very little in the form of protection.

Parasaurolophus is an easily recognizable dino and features in many dinosaur books and toy ranges today.

# UTAHRAPTOR

Utahraptor, meaning "Utah Robber", is another of those small, fast, carnivorous dinosaurs of the dromaeosaurid family, which includes Velociraptor and Deinonychus. Remember those long slashing claws on the hind foot of Velociraptor? Well, Utahraptor is one of those kinds of dinosaurs. It is the largest known member of the family – about the size of the so-called Velociraptors which we saw in Jurassic Park.

Utahraptor lived during the Early Cretaceous Period, about 125 million years ago in what is now Utah, USA. Utahraptor is the largest known dromaeosaur.

It is twice the size of Deinonychus. Perhaps one of the deadliest of all dinosaur predators. Utahraptor had a huge, sickle-shaped slashing talon on each foot and sharp claws on its hands. When fully grown, its deadly claw would grow to about a foot in length. Utahraptor is believed to have been an agile predator. It possessed powerful legs, a long stiff tail which aids the dinosaur to maintain its balance while chasing its prey. Utahraptor would use this speed in combination with brute force to bring down its prey. Its huge tail was sure to inflict deep wounds by slashing out with its feet. Utahraptor would jump on its victim's backs, once on the back, it will hold on with its grasping hands and sharp serrated flesh cutting teeth. It will use its killer claws to tear into an unprotected part of its victim's body, such as its belly.

The first Utahraptor specimen ever found comprises skull fragments, a tibia, claws and some tail vertebrae from which it has been inferred to be twice the length of Deinonychus. Those slashing claws on its second toe were up to 8,5 inches in length. Go and grab your ruler and see really how long those claws were. From the largest specimen found, Utahraptor is estimated to have reached a length of 23 feet and weighed in at half a ton.

There is strong evidence that Utahraptor was feathered based on the fact that Velociraptor has been found with feathers and being from the same family.

Utahraptor was found in 1975 by Jim Jensen in east-central Utah, close to the town of Moab but for some reason, it didn't get much attention. In 1991 a large foot claw was found by Carl Limoni and then the paleontologists did get involved. Kirkland, Gaston, and Burge dug up some more remains in Gaston Quarry in Grand County within the Cedar Mountain Formation. Age dating has shown that these rocks were laid down about 124 million years ago. The type specimen is housed at the College of Eastern Utah Prehistoric Museum whilst the Brigham Young University houses the largest collection of Utahraptor fossils.

Utahraptor ostrommaysorum the type species was named after John Ostrom of the Peabody Museum of Natural History and Chris Mays of Dinamation International. They nearly named it spielbergi after the film director Steven Spielberg.

In 2001, six individual Utahraptors were found comprising an adult, four juveniles, and a baby, along with the bones of an Iguanodon. It is believed that the Utahraptors were attacking the Iguanodon which was stuck in thick mud or quicksand and themselves caught in the same trap as they preyed on their helpless victim. There have been other examples of predators being caught most notably the La Brea Tar Pits. If the theory is true it would imply that Utahraptors hunted in packs which provides a fantastic insight into their behavior.

# GIGANOTOSAURUS

As the name simply implies, Giganotosaurus was a very large meat-eating dinosaur, which have equaled the likes of Tyrannosaurus Rex in build and height, with fossil remains proving that in some cases they were in fact slightly larger. Estimates have placed the length of the giant dinosaur, Giganotosaurus, to be around the 46-foot mark, and it would have walked upright on two legs. . Despite the imposing name of Giganotosaurus, we do know there were larger meat-eaters that have existed, such as the Spinosaurus, which is perhaps one of the reasons why they have no same name recognition as the aforementioned dinosaurs.

Giganotosaurus or Giga as we are now going to refer to them would have lived in the late Cretaceous period, around 100 million to 96 million years ago. Fossil remains have been discovered in Argentina, a part of South America which is becoming a real hotbed for finding larger and larger dinosaur remains.

The first discovery of the Giga was in the Patagonia region of Argentina in 1994, and the fossils were found close to the remains of a 75-foot long plant eater, which has left some experts speculating that it had fallen victim to the Giga. The dinosaur gained its name in 1995 and it maybe this fact, that they have only been known for a short spell, which is why it isn't as instantly recognizable as some of the other more famous dinosaurs that were found in the 19th century.

Gigas would potentially have weighed in at around the 8-ton mark, which is slightly lighter than the Tyrannosaurus Rex. A fully grown adult would have stood over 11 feet at the hip and have large serrated teeth, roughly 8 inches long. Similar to all dinosaurs with this shape, tall with small arms and a large head, there has been a debate in relation to the speed of travel. Again, similar to the T-Rex, experts have concluded that Giga would have been very fast using its tail as a counterbalance and would have hunted down its prey, rather than being a scavenger.

Giganotosaurus is yet another amazing dinosaur. The fact that we discovered this creature is probably the main reason why most individuals haven't heard of them. If fortunes were reversed and it had been discovered in the early years of fossil collecting, then Giga would probably have featured in many films and be a household name similar to the Tyrannosaurus Rex. There are some great images of how experts believe this dinosaur looked, which can be found in dinosaur books and online.

# IGUANODON

Iguanodon lived during the Early Cretaceous Period, about 110 million years ago. Iguanodon fossils have been discovered in South Dakota, Utah, USA, and in Europe. Iguanodon was the largest member of its family. It could stand three times the height of a man and weight up to 4.5 tons. Iguanodon is one of the most famous dinosaurs and a more primitive relative of the duck-billed hadrosaurs, because it was discovered in 1822, by Mary Mantell, when dinosaurs were still unknown to science. Iguanodon was the second dinosaur discovered and identified by scientists - in fact before the word "dinosaur" had even been invented - the first fossils of the animal were found in England in 1809. Since 1809, fossils of Iguanodon have also been found in a number of other countries around the world including Belgium, Germany, the United States, Mongolia, as well as in Africa.

Additionally, footprints, but no skeletons has yet been found in Spitzbergen Island (which is in the Arctic Circle) and South America.

Iguanodon probably mostly walked in a bipedal (two legged) posture, also possibly able to stand on four legs. Its most unusual feature is the conical horn attached to each thumb - the purpose of this horn is not entirely clear - although defense or obtaining foods are likely possibilities. In terms of size, Iguanodon was about 30 feet (9.1 meters) long.

Iguanodon is a good plant eater with is long skull, beaklike jaws, and rows of grinding cheek teeth that help to distinguish ornithopods from other herbivorous dinosaurs. The impressive closely packed battery of teeth made it well-suitable for Iguanodon in grinding tough plant material. The upper surface of each tooth was broad and ridged.

The first reconstruction of Iguanodon was made by Gideon Mantell in the 1820s. Mantall only had limited information and partial skeletons to work from, so he did make a number of errors, including wrongly attaching the horn to Iguanodon's snout.

Another famous early reconstruction of Iguanodon was made by the sculptor, Benjamin Waterhouse Hawkin, with advice from Sir Richard Owen. This sculpture attempted to depict Iguanodon as it as always been in life (but wrongly depicted it in a sprawling lizard-like posture, and with a horn on its snout). The sculpture is perhaps most famous for being the site of a dinner party that Owen held on New Year's Eve 1853 for 21 eminent men of science. Today, this sculpture (and a variety of others constructed during the same period) can be found in Dinosaur Court in South London.

Study of Iguanodon specimens showed that, far from being heavy, lumbering quadruped. Iguanodon was relatively light in contrast to its great length and could move on its hindlimbs. Studies of its front legs suggest that Iguanodon had strong legs and that the central three fingers ended in hooves. This implied that Iguanodon spent some time on all fours, and probably ran on its hindlimbs when it needed to move quickly.

# DIPLODOCUS

Diplodocus was one of the largest dinosaurs ever discovered. It was the longest land animal, but not the heaviest. It was a sauropod, a "lizard-hipped" dinosaur. Diplodocus was a plant eater and had a long neck reaching around 26 feet, but it could not hold its neck beyond 17 feet off the ground. It had a very long tail (around 46 feet) that was whip-like in appearance. Diplodocus had a head, with nostrils on top that was a little less than 2 feet long. The Diplodocus' had significant short front legs when compared to its back ones. Its legs featured feet with five toes each, very elephant-like in appearance. Each foot had one thumb claw, which is believed to have been a protection mechanism.

Most of the length of this herbivorous animal was in the neck and tail. Diplodocus had an elongated snout with nostrils on top of the head and simple blunt peg-like teeth that were useful for nipping at and stripping foliage. It had one of the smallest brains (the size of a fist), its intelligence was among the lowest of the dinosaurs. The dinosaur was 90 feet long and weighed between 10 and 20 tons. It lived about 155 million years ago.

Diplodocus carnegii was named after industrialist and philanthropist Andrew Carnegie. This sauropod was the most famous dinosaur on the planet in the last century. There is a famous life-size replica of the Diplodocus nicknamed "Dippy" in front of the Carnegie Museum of National History (Pittsburgh, PA, USA). Diplodocus was a herbivore, it ate only plants; main food was probably conifers. It must have eaten an enormous amount of plants each day. It didn't chew them, it swallowed leaves whole. It swallowed also stones (gastroliths), which it held in its stomach. They helped by digesting tough plant material. Diplodocus may have traveled in herds, migrating when the food supply was depleted. It probably bred from eggs, like other sauropods.

It is thought, that the eggs were laid as the dinosaur was walking. Sauropods such as Diplodocus were believed to have dual brains. Later, it was discovered that this so-called second brain was nothing more than the spinal cord that had an extension/ enlargement. This enlargement, found in the hip area was even larger than the Diplodocus' already small brain.

The name Diplodocus means "double beam" and describes a special feature of the backbone. It had slender limbs and hind legs, longer than the front legs; it gave him access to both low and high growing plants. It was quadrupedal, it moved slowly on four column-like legs.

# THE FASCINATING HISTORY
# OF DINOSAURS

It seems almost impossible for any child to come through school without developing at least a passing interest in Dinosaur Trivia. Dinosaurs were incredible beings by the standards we know today and were clearly the dominant species on the planet for millions and millions of years. The periods of time when the Dinosaurs roamed the Earth are known as the Triassic, Jurassic, and Cretaceous period.

Any dinosaur quiz will usually show that there were many different types of dinosaurs. Scientists have discovered over 1000 different breeds of the creatures, and believe it is possible that they have only found a fraction of those that actually existed. Most of them were wiped out in an extinction event that took place approximately 65 million years ago. Moreover, scientists do believe that dinosaurs have descendants roaming the Earth today. It has become commonly accepted scientific cannon that birds evolved from certain types of dinosaurs.

Dinosaur trivia will show that the appearance of the reptiles varied greatly, as can be seen in the diverse skeletons which have been reassembled in museums around the world. There are many varieties of dinosaur which are distinctive based on the shape of their skeletons.

For instance, the stegosaurus had large, triangular plates protruding from its spine. The triceratops had three horns emerging from its skull plate. The Tyrannosaurus Rex, which is one of the most well-known dinosaurs was a fearsome beast that walked upright on two legs, with two tiny upper arms.

A Dinosaur quiz will show that the size of these creatures could vary as much as their appearance. Although popular culture brings to mind huge monsters with the word dinosaur, in truth, some were the same size as humans, and even as small as modern rodents. On the other hand, there were other varieties that were much larger.

To get an idea of the size of the larger dinosaur, look at the Tyrannosaurs, one of the larger predators. It could grow up to thirteen feet tall, and 43 feet long, weighing over seven tons. There were plant-eating dinosaurs which grew to be much larger even than this. The largest of these families were known as the Sauropods.

The Brachiosaurus is one of the largest of the sauropod family and was one of the largest things which have ever lived. These dinosaurs could have reached over 80 feet in length, and could probably lift their head over 40 feet into the air.

Truly, birds are living dinosaurs and a glorious collection of fossils makes this feathered revolution abundantly clear, but a chickadee is no stand-in for a Dreadnoughtus. And when we look at the bones of the extinct, non-avian dinosaurs that haunt our imagination, we're faced with a seemingly infinite number of questions about how they lived. What did they eat? How fast could they move? What did they sound like? What colour were they? It's impossible for me to look at a dinosaur, reconstructed or in the raw, and not wonder about the life of that animal, and I believe that this basic sense of wonder about the lives of the bizarre and the extinct accounts for dinosaurs' powerful allure.

There is an even deeper dinosaurian connection, though. Whether it's noting the time a specific dinosaur lived, or comparing their size to a double-decker bus – the popular standard for dinosaurian measurement – we're constantly considering dinosaurs in relation to ourselves and the world we know. **Dinosaurs** embody the drastic changes that life on Earth has undergone, and give us access to some of the most powerful truths our species has come to understand – that our planet has an incredibly deep history, that life has changed constantly through time, and that extinction is the fate of all.

Chasing after dinosaurs is really a quest to fill in part of our own backstory, not least of all because our own fuzzy, shrew-like ancestors scurried under their feet for over 160 million years. Dinosaurs can be Hollywood monsters, objects of scientific fascination and everything in between, but at the root of it, our fascination with them stems from wanting to know more about the prehistory we share. The dinosaur story is part of our own.

# BONUS CHAPTER: THE LOST ISLAND (SHORT STORY)

Dinosaurs were extinct. The only evidence of Dinosaur existence is the discovery of fossils by the Paleontologists Unearthing small details about these larger than life characters that once walked upon this earth before the world as we knew it ever existed. Before large skyscraper buildings, motor vehicles and our sociological world ever existed. Dinosaurs lived a much simpler life. It was unfortunate that Dinosaurs went extinct sixty-five million years ago.

What if that wasn't true? What if the World's most influential people had kept a secret? The secret and extraordinary mind-boggling lie everyone will hear and laugh, thinking is nothing more than an insane conspiracy theory.

Perhaps that is why a group of expeditioners did years of extensive research, interviewing and searching several different countries and continents. After years of trying and failure, the group of expeditioners and paleontologists almost began to believe that perhaps the theory that dinosaurs were still very much alive was nothing more than a conspiracy. That was until one day...

The group had been given a hint from an unknown source about a secret and very discreet location. It does not appear on any map, there was no slight trace of evidence that the place existed. But it was very much real. The group of expeditioners and paleontologists stepped off their boat onto the soft moist sand that formed around their feet. It appeared to be a regular island, a dense thick bush only meters away from the ocean. While the Island appeared to be something out of a vacation brochure, the nerves that radiated off each member of the team solidified that there was something about this place. Curiosity had been sparked. Adrenaline was pumping. Would their determination finally pay off? The team didn't have to wait long to discover if their trip had been wasted. A loud, terrifying, bone-shaking roar swept through the trees as they trembled caused by the vibrations. Something knew that the team had arrived. Whatever it was, certainly wasn't happy.

The team sprang into action. Everything had led up to this moment. With their high-tech scientific equipment, the team left the beach determined to break into the forest to discover what can make such a horrifying noise. The team grew deeper and deeper, passed the winding trees and tangling bushes when suddenly they stopped. Not a sound was made. In front of them was a view that people could dream off. Shades of green and browns protected the utopia that cascaded before them. As the forest became less dense, the most magnificent and undeniable exhibition of earth's mightiest creations laid out before them. A group of Pterosaurs flew graciously in the air searching for their next meal. Creatures invisible to human sight chirped, grunted trying to contact with their family. Not too far off in the distance, a female Triceratops with her young calf fed mindlessly on deep rich green leaves.

Yet only too soon did the euphoria of the occasion wear off and human's nature to conquer returned. Like a pirate hunting for treasure, the need to capture and expose the world's greatest secret resume. The team clubbed together forming a magnificent and full proof plan. Or so they thought. The leader Thomas O'Reilly, world renowned paleontologist thought he knew these creatures he had studied for the past twenty years. But nothing could have prepared him for the battle he faced.

The team attempted to approach the female and her calf from behind with great confidence. Stepping closer and closer, not expecting the female Triceratops to swiftly turn around lowering her head before charging at the team who believed that she could not see nor hear them. The charge was successful, knocking several team members out of the way. None of them were seriously injured because that wasn't her intent. While the team members recovered from the ambush, she remained focused on the group, watching them intently. One foolish member attempted to go for a solo attack, yet again she reacted instantly.

Thomas O'Reilly was dumbfounded and confused. Triceratops was never reported to be such intelligent creatures with extensive senses. The team attempted several different tactics but she seemed to counteract each of them with a spectacular ease. Thomas O'Reilly didn't want to admit defeat, but at the end he did. They could find another creature that was perhaps easier to control.

However, as the team explored the unknown habitat full of earth's dangerous and extraordinary creatures. The creatures seemed to be more aware of their presence. A herd of Brachiosaurus were quick to slowly form a tight circle around their elderly and the young. Velociraptors communicated deeper within the bushes, unable to be seen via the human eye, yet the Velociraptors knew exactly where the humans were, in fact they were keeping an eye on them.

The team was too preoccupied with the noises and bush rustling when it suddenly stopped. A large shadow cast over the team. The youngest member of the team looked up gasping. In unison, the team all looked up to see the King of the Dinosaurs, Tyrannosaurus Rex. They thought it was wise to attempt to stand their ground. But the Tyrannosaurus Rex was smart, with one bone-shaking roar, the team instinctively began to run for their lives, leaping over overgrown stumps and winding plants. This was one animal that wasn't to be captured. Instead, it was trying to capture them.

The team was chased for the several miles that they had covered throughout their expedition. Their feet began to sink into the soft sand. They quickly jumped back onto their boat away from the island. Suddenly it became clear to them. If the dinosaurs had managed to survive extinction, the Earth may be a very different story. Just like humans, dinosaurs would find a way to adapt and keep themselves on the top of a very large and complicated food chain. If Dinosaurs existed to this very day, they would very much indeed rule the world.

# CONCLUSION

Thank you again for purchasing this book, I hope you have enjoyed it!

Can I ask you a favor? If you did enjoy this book, kindly review on Amazon. If you search for my name and the title on Amazon you will find it. Thank you so much, it is very much appreciated!

*George Owen*

Made in the USA
Middletown, DE
13 July 2019